Wartime ploughing of pasture, Carnforth, Lancashire, 1943.

PLOUGHS AND PLOUGHING

Roy Brigden

Shire Publications Ltd

CONTENTS

Introduction 3
The early plough 7
Ploughs of the eighteenth and nineteenth centuries 11
Plough operation 17
Mechanical ploughing 25
Places to visit 31
Further reading 32

Published in 1998 by Shire Publications Ltd, Cromwell House, Church Street, Princes Risborough, Buckinghamshire HP27 9AA, UK.

First published 1984; reprinted 1998. Shire Album 125. ISBN 0 85263 695 4.

Printed in Great Britain by CIT Printing Services, Press Buildings, Merlins Bridge, Haverfordwest, Pembrokeshire SA61 1XF.

Brigden, Roy
Ploughs and Ploughing.–(Shire albums; 125)
1. Plows–History
I. Title
631.3'12'09 S683

ISBN 0-85263-695-4

ACKNOWLEDGEMENTS

The author acknowledges with gratitude the advice frequently provided by David Phillips, Archivist, Institute of Agricultural History. The drawings are by Godfrey Eke of the Museum of English Rural Life. All photographs (except the front cover) are the copyright of the Rural History Centre, University of Reading. The cover photograph is reproduced by permission of the Victoria and Albert Museum, London.

COVER: *Ploughing near Scarva, County Down, 1791. Print by William Hincks.*

Naming the parts of the early twentieth-century horse plough.

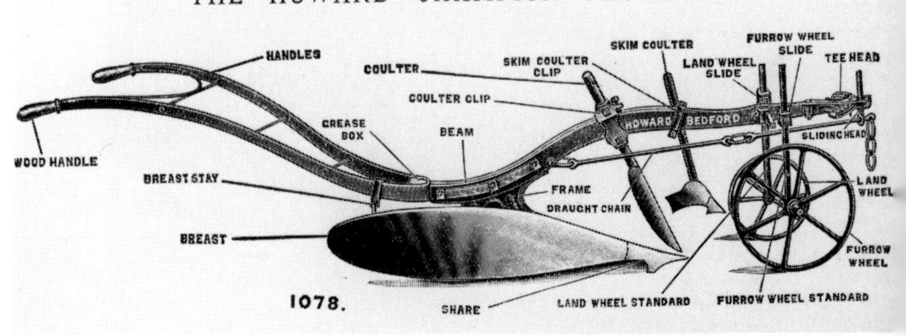

immy and Joey, the last working oxen in Britain, photographed in Earl Bathurst's park at irencester, Gloucestershire, 1957.

INTRODUCTION

loughing is perhaps the most basic of all he operations upon which systematic arming depends. Notwithstanding more ecent and revolutionary methods of seed ed preparation, the plough is still one of he most immediate images of the con-emporary countryside. Its work of today vould be recognisable to a medieval armer even though the understanding, he technology and the motive power ave changed much over the intervening enturies.

Seeds cannot germinate effectively un-ess prior tillage by man has formed a rotective and nourishing tilth; the young lants cannot grow and prosper if they ave to compete with a choking profu-ion of weed. The act of ploughing is part f the preparatory process vital for crop levelopment as well as for the continuing naintenance and conditioning of agri-ultural land. A conventional plough, of ny period, uses a *coulter* and *share* to cut soil slice, which is then pushed to one ide and inverted by the *mouldboard.* These, the three essential working parts of the plough, are connected via the main beam to the source of traction.

Inversion of the top few inches of soil has the immediate benefit of burying the vegetative matter it contains and thereby speeding its decay. This applies to weeds and the remains of previous crops but manure may also be spread on the surface and ploughed in so that subse-quent chemical action releases its nut-rients. At the same time, turning a furrow slice brings to the surface soil that has lain buried and undisturbed for many months beneath the previous crop. Here, it is now subjected to the weathering proces-ses of frost, wind and rain, which com-bine to break down the consolidated mass into a crumbly texture. As a result, exposure at the surface allows air to permeate through the soil structure and provide the conditions necessary for be-neficial bacterial activity. This process of aeration is one of the prime reasons for ploughing for it ensures a fine, sweet-

Hampshire plough by William Tasker and Sons of the Waterloo Ironworks, Andover (MERL collection). Following local practice this is a gallows plough and is fitted with a screw head. It was made around 1892 and was in use on a farm at Middle Wallop, near Stockbridge, until the early 1920s.

smelling tilth at seed time.

Ploughing also has a part to play in moisture control. With the furrow slices laid up in ridges, one against the other, the winter rains do not remain for long periods on the surface, where they cause waterlogging, but rather are encouraged to penetrate quickly through to the subsoil. A store of moisture is so created which can slowly rise up through the layers, after the seed bed has been firmed, to sustain the plants throughout the growing season.

A full scientific explanation of the several functions of the plough did not come until the twentieth century. But long before this an awareness of its fundamental importance had found expression in a wider symbolism. Ancient ploughs, for example, were not infrequently a subject for the cave painters and, certainly in parts of Europe, also featured in ritual burials. Daily life in Anglo-Saxon England bore the permanent mark of the plough, for the amount of land that one ox team could plough in a normal year was used as a unit of measurement and means of tax assessment. The area was termed a hide and extended to about 120 acres (50 ha). Even the acre itself may have originated as the capacity of one plough in a day.

There was also a place in the annual calendar of festivals. Plough Monday fe on the first Monday after the twelfth da of Christmas and marked the beginnin of the new ploughing season in readines for the planting of spring corn. In north ern and eastern Britain especially, th day was celebrated by ploughmen an boys in fancy dress hauling a plough fron door to door and collecting money for th festivities. Any householder well able bu unwilling to pay might find a deep furro cut outside his front door.

The ability to plough a field well ha always been regarded as more of an ar than a mere skill. As well as the techni que and experience, the stamina an doggedness, and the mastery over beas or machine, there had to be an infusion natural, indefinable ability. This was th quality that created local heroes at th popular ploughing matches of th nineteenth century, when neighbourin rivalries whipped-up much spirited com petition. It was also the quality tha raised the ploughman and his team abov the common anonymity of rural society confirming a degree of agricultural nobi ity upon both. A good ploughman wa worth the extra wages he received. A badly ploughed field could not be hidde but remained on public view for all to se and scorn.

.BOVE: *Local ploughing match at Shute Cross, Devon, 1940s.*

ꞋELOW: *A well earned rest for horse and man.*

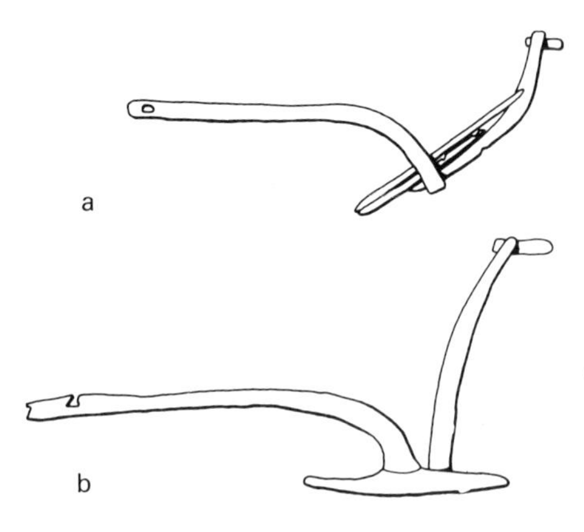

Reconstructions of early wooden ards disco vered in peat bogs in Denmark: (a) is a bow ar from Donnerupland with the oak guidin handle, main share and fore-share mortise through a hole in the beam. This allows th share and fore-share to enter the soil at a angle. (b) is a crook ard from Vebbestrup. A forked piece of timber, in this case birch, form the beam and sole into which a handle or stilt i mortised. The point of the ard travels throug the soil at a more nearly horizontal attitude Although no complete wooden ards from late prehistoric Britain have been found, the evi dence available suggests that though both type were in use the bow ard was more widespread.

A wheeled plough depicted in the Gorleston Psalter of c1310-25. Manuscript illustrations canno always be relied upon for accuracy because often they were themselves copies made by an artist wit little or no knowledge of ploughs. This one is interesting in that it shows a mixed team of a horse an a pair of oxen.

An English swing plough from the Luttrell Psalter of about 1340. This is a very good illustration of a medieval plough built on the rectangular frame principle with dipping beam, coulter and lon wooden mouldboard. A second person, often a boy, to goad the beasts on was a necessary feature o ox ploughing. Ploughs not dissimilar to this were still being made in the seventeenth century.

Kentish turn-wrest plough from 'General View of the Agriculture of the County of Kent' by John Boys, 1805. Amongst the agricultural writers of the day who criticised the heavy, lumbering appearance of this plough was William Marshall. Even he, however, accepted that for steep downland surfaces and absorbent subsoils it was 'obviously and admirably adapted'. In Boys's description the plough has a beam up to 10 feet (3 m) long and a share 4 to 7 inches (114-178 mm) wide. Other ploughs had been tried in the area, he says, but none were as well suited to local conditions.

THE EARLY PLOUGH

Knowledge of ploughs and ploughing methods used in Britain before the arrival of the Romans has to be pieced together from the necessarily incomplete evidence so far discovered. The earliest visible indication of ploughing comes in the form of scratch marks made by the point of the share as it penetrated through to the subsoil at a greater depth than normal. Such marks, dating from around 2800 BC, have been found in the course of excavations at Avebury in Wiltshire. Other examples from the continent of Europe precede this by perhaps eight hundred years.

The implement most likely responsible, having neither coulter nor mouldboard, is more accurately termed an *ard* than a plough. Peat bogs in Denmark have yielded some well preserved specimens, the earliest made around 1500 BC, which have been grouped into two principal types. The *bow ard* consists of a curving beam with an angled share and a guide handle, or stilt, mortised through a hole at one end. On the other hand, the *crook ard* has both main beam and share beam composed of a single piece of timber with the stilt forming a separate fixture. Parts of ards have been found in Britain, mostly of the bow type but also of the crook variety. Although the ear-

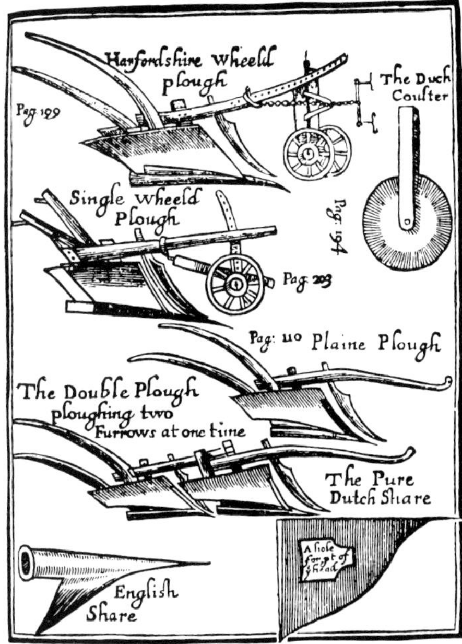

Seventeenth-century ploughs, including the Hertfordshire, from Walter Blith's 'English Improver Improved' of 1649. Blith opened up the literary debate on plough design and, although he left many questions unanswered, he did make some suggestions, most notably with regard to the shape of the mouldboard.

Mid nineteenth-century Kent plough used in the Folkestone area (MERL collection). Rather than laying up the soil in seams, the Kent plough completely inverted it to leave a flat, crumbly bed. This was reckoned to be an advantage when the soil was of a naturally dry character. At each turn, the detachable iron mouldboard was fixed to the other side and the coulter position shifted by altering the wedging stick on top of the beam. At the end of the nineteenth century, these ploughs were still employed by the majority of farmers in mid Kent and the Weald.

liest of these date only from the last few centuries BC, the identification in the Shetlands of many stone share points of around 1650 BC confirms that ard cultivation in Britain dates back much further.

With such basic features, the ard could do little more than penetrate and stir the land. Signs of uneven wear on the surviving shares and fore-shares, however, suggests that the implement was tilted in use to one side, usually the right, in order to push the soil laterally and form a crude furrow. It is likely that more effective soil disturbance was achieved through *cross ploughing,* in which at each ploughing the field was gone over in two directions at right angles to each other. This method might help to account for the characteristic square or near square shape of early fields.

Some iron parts for ards, notably share tips, have been discovered from later pre-Roman times in Britain but evidence for the use of iron becomes much more common during the Roman period. It is clear that the ard was still in use over the first few centuries AD but the existence in some numbers both of large iron coulters, weighing anything up to 16 pounds (7 kg) each, and iron shares of Romano-British origin suggests the cutting of a furrow slice and therefore implies that a much heavier plough equipped with a mouldboard had already been introduced. Possibly climatic changes influenced this. As Britain grew colder and damper during the early iron age, so the light cultivation methods associated with the need to conserve soil moisture became less appropriate. Hence the mouldboard plough, capable of cutting and turning a wet furrow slice to assist with drainage and weathering, was developed.

With the heavier implement, working at greater depth, cross ploughing was no longer necessary. As a result, there is some indication even from Roman times of a change in field shape to longer strips having a length four or five times greater than the breadth. Through the medieval period this pattern was accentuated because ploughing in long thin units reduced to a minimum the amount of time and space wasted in turning the implement at each end. Not infrequently, strips in the open fields developed an elongated S formation in order further to assist the turning at the headlands of the often cumbersome plough drawn by its team of four or more oxen.

Hard evidence on the design and construction of medieval ploughs is scarce. From illustrations in early manuscripts of both wheeled and wheel-less or swing ploughs, it seems likely that the process by which ploughs evolved differently around Britain, according to a combination of external influence and local needs, was already under way.

More information survives for ploughs of the sixteenth and seventeenth centuries because this was a period of considerable literary debate on the current state of agricultural practice. By now the regional variations of design were becoming marked. What was known as the *Hertfordshire plough,* widely used in the Midlands, may have been a descendant of the Saxon wheeled plough. It was very heavy with wheels 18 or more inches (460 mm) in diameter and a long main beam 6 feet (1830 mm) in length. The share beam carried a long pointed iron share and was attached to the beam via a wooden sheath and a land or left handle. The flat wooden mouldboard was similarly fixed to the sheath and right handle.

In *Kent* a double-wheeled plough of ancient origin and massive proportions was in common use, characterised above all by its ability to turn furrows to either the left or right side. Alternatively in *East Anglia* a light plough of some Dutch origin had been introduced towards the end of the sixteenth century; it could be drawn by two horses, rather than four oxen, and with one man could work 2 acres (0.8 ha) of thin Norfolk or Suffolk soil in a day. In spite of later developments in plough theory and design, recognisable forms of some of these earlier types, the Kent plough in particular, were still being built and used in the nineteenth century.

Ploughing, scenes by W. H. Pyne. First published 1802, they later appeared in the 'Microcosm' of 1806. Pyne was critical of the use, especially in the south and west, of immense unwieldy ploughs and team of four horses: 'as much land would be ploughed with a light plough drawn by two stout well-fed horses, moving on spiritedly under the management of the ploughman alone, as by the great lumbering thing of our forefathers, drawn by four with an additional lazy boy doing nothing but walking by the side of the two fore ones, while all six advance with the old immemorial sluggish step, and stop at every dozen yards.'

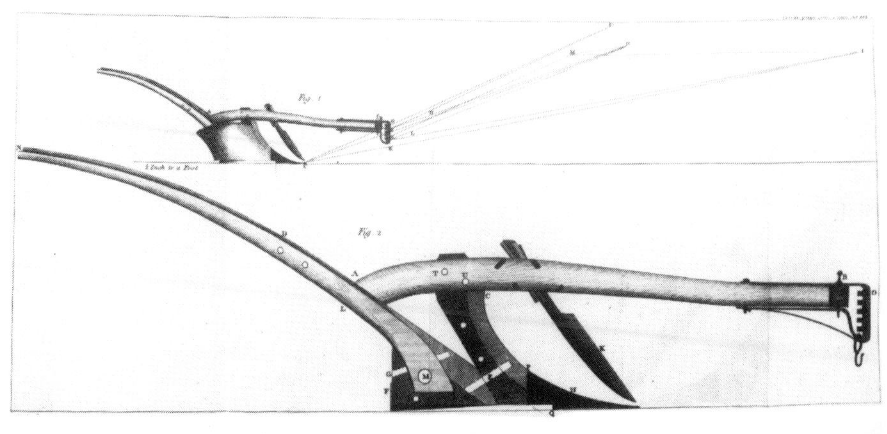

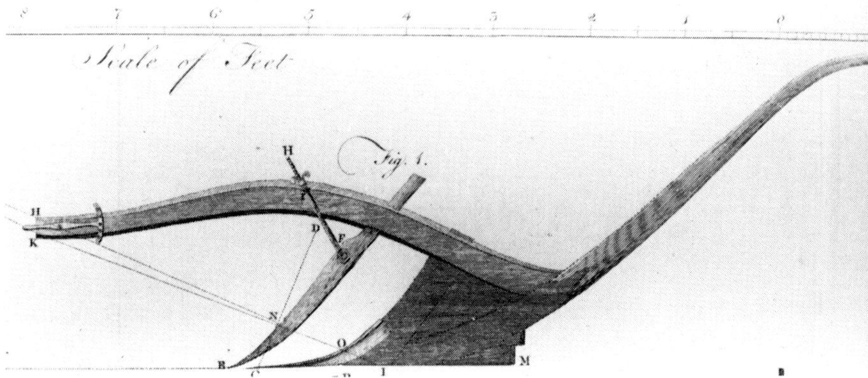

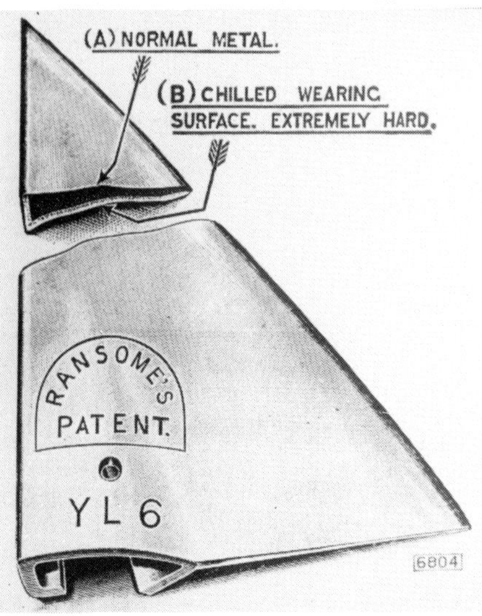

TOP: *Illustration from John Arbuthnot's article on ploughs, in Arthur Young's 'Farmer's Tour through the East of England', Volume II, 1771. Arbuthnot made a number of improvements on the basic Rotherham design including a broader share rising into a breast that followed the shape of a quarter ellipse. He achieved a desirable curve on the mouldboard by fashioning it from wood, allowing it to be worn further by the action of the soil, and then plating it with iron.*

ABOVE: *A swing plough of the Rotherham type from James Small's 'Treatise on Ploughs and Wheel Carriages', 1784. Small was himself a ploughwright who popularised the Rotherham plough in Scotland and made important contributions to developing the theory of plough design, notably with his detailed analysis of the shape of mouldboards.*

LEFT: *The composition of a Ransome chilled cast iron share. The concept of chilling cast iron, embodied in Ransome's 1803 patent, is still operable in the twentieth century.*

A Ransome's 'A' Plough with single handle, c1850 (MERL collection). This was a general purpose plough popular in the eastern counties to which fifty varieties of mouldboard could be fixed. Price £2 11s 6d.

PLOUGHS OF THE EIGHTEENTH AND NINETEENTH CENTURIES

Important developments in the three related areas of plough theory, design and construction came during the eighteenth century. Critical comment had already been mounting against the heavy and unwieldy characteristics of many ploughs in use. It was not so much a problem of weight alone but rather that the draught power requirement was increased by the various compensatory devices needed to counteract more fundamental inadequacies in design.

All the accepted conventional wisdom concerning the plough was rethought. Horses were coming into wider use for draught purposes and the East Anglian plough had demonstrated the viability of a light, fast implement drawn by only two horses and dispensing with a driver to lead the team.

A significant step forward occurred in 1730 when a patent was granted to a new plough that was to become known as the *Rotherham,* after the town of its origin. It was developed by Joseph Foljambe in collaboration with a Sheffield businessman, Disney Stanyforth, who provided the financial backing. Of the plough's several novel features, perhaps the most important was the absence of a share beam. This unwieldy component, dragging along the floor of the furrow, had added a great deal of weight to the plough and restricted it to the standard rectangular frame configuration. The Rotherham, by contrast, had only a heel at its base and the land handle slanted forward towards the breast to form a triangular frame. This simpler arrangement was not only less cumbersome but also more rigid, thereby allowing a lighter form of construction to be employed. Without a share beam, the breast and share became more of a unit flowing on naturally into the curve of the mouldboard. The result was a faster plough, of less draught, over which the ploughman could exert more control by reason of the long handles and greater leverage.

The Rotherham plough did not lead to any immediate revolution but it pointed the way for the future and its principles were commonly reflected in ploughs of a century later. It provided the starting point from which commentators and manufacturers could further explore the theory and practice of plough design. In their writings in the second half of the eighteenth century, for example, both James Small, a Scottish ploughwright, and John Arbuthnot, a Norfolk farmer, sought to refine the Rotherham plough and devise mathematical models for the most effective shape of the breast and above all of the mouldboard.

It was not, however, until the new

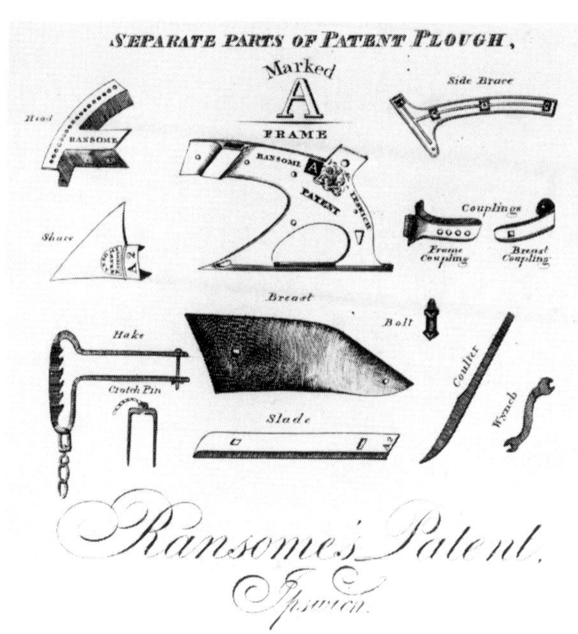

Information sheet, c1830, showing the iron components of the Ransome patent 'A' plough. This was given to customers so that, should replacement parts be required later, they could be ordered by name without risk of misunderstanding.

theories were aligned with new materials and methods of manufacture that plough development began to accelerate. This was essentially a phenomenon of the nineteenth century but again the Rotherham laid the foundations. Under Foljambe's direction it was factory built in Yorkshire, at a potential output of around three hundred per year, from standardised wooden parts made from sets of master patterns. Volume production was to be the key because the mere existence of an improved design was not enough to change long established practices. The new ploughs had to be available in sufficient numbers to compete with, and slowly oust, the individually made idiosyncratic creations of the village workshop.

Factory production of ploughs is associated above all with the name of Robert Ransome. Born in Norfolk in 1753, he set out as an ironmonger and founder in Norwich but by 1789 had moved to Ipswich to establish what was to become one of the most noted agricultural implement and machinery works in the world. Cast iron was the ideal material of the day to use in large-scale output because any number of identical components could be made from moulds formed by the original pattern. It had the great disadvantage, however, of being brittle and therefore far less suitable than wrought iron for an implement such as the plough which has to be dragged through the soil.

Ransome was not the first to make plough parts out of cast iron but he was responsible for important improvements to the quality of the material. These were covered by two patents, one in 1785 for tempering cast iron plough shares with salt water and the other in 1803 for selective 'chilling' or hardening of cast iron. The result was that the plough share could now be successfully reproduced by the thousand. Moreover, the Ransome shares were hardened on the underside so that the softer upper surface wore away more quickly and automatically maintained the sharpness of the cutting edge.

Even before the end of the eighteenth century Ransome was distributing his cast iron shares across Norfolk and Suffolk through appointed agents. Through them, farmers could reorder shares to their exact requirements simply by quoting the relevant pattern number. At the Ipswich works the implements were

HOWARD'S CATALOGUE, 1878. 11

ILLUSTRATIONS OF HOWARD'S CHAMPION PLOUGH AS VARIOUSLY FITTED.

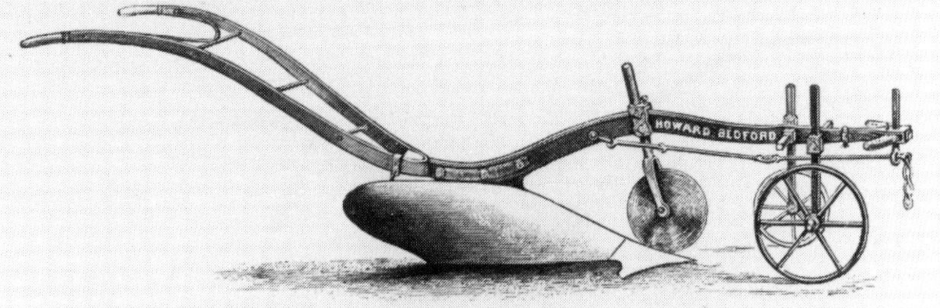

The General Purpose Plough—Mark B—fitted with Skeith or Revolving Coulter, for FEN or MOOR LAND.

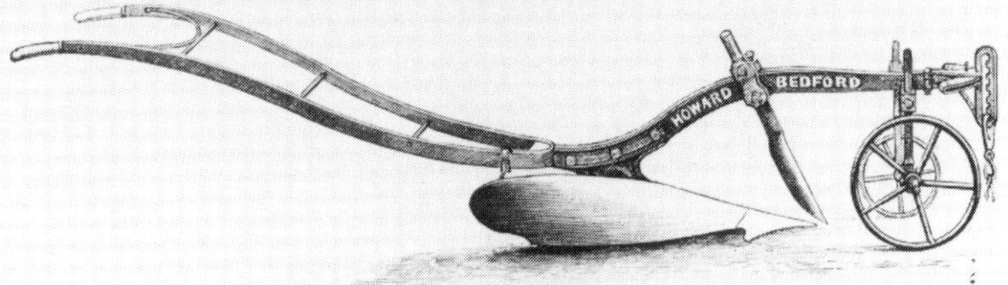

The B Plough fitted with Breast and Share for HIGH CUTTING or TRAPEZOIDAL SHAPED FURROWS.

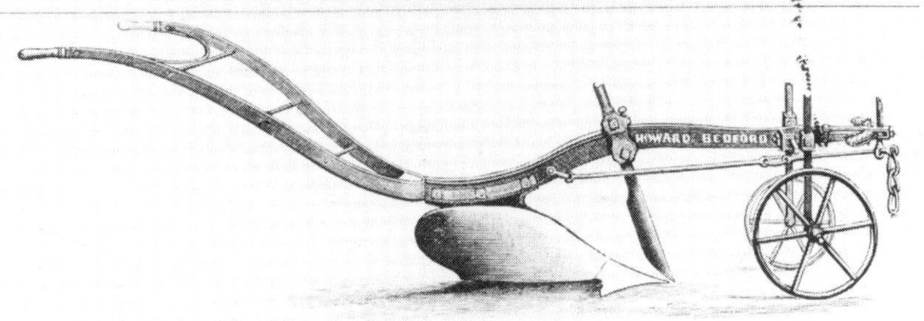

The B Plough fitted with SHORT BREAST.

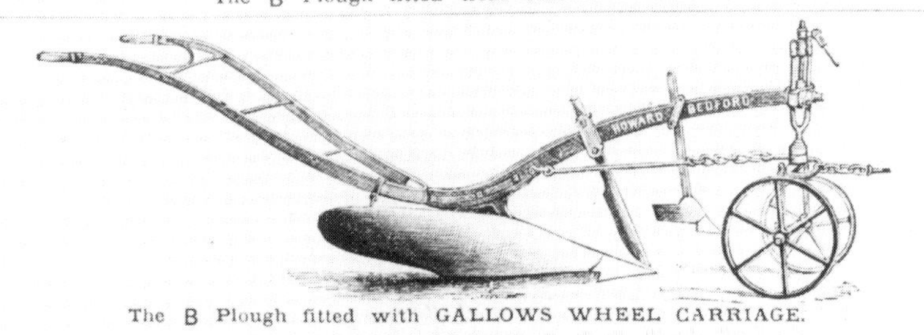

The B Plough fitted with GALLOWS WHEEL CARRIAGE.

James Howard followed his father into implement manufacture and at the age of nineteen won the first prize for an all-iron plough of his own design at the 1840 show of the Royal Agricultural Society. This was the forerunner of the Howard Champion series of ploughs which boosted the fortunes of the family firm and led to an international reputation. Many variants of the Champion plough were available, according to district and type of work.

ABOVE: *Double-furrow horse plough by R. Hornsby and Sons of the Spittlegate Ironworks, Grantham, 1890. This was one of a range of similar Hornsby ploughs fitted with central lifting apparatus to raise the plough clear at the headlands and facilitate turning.*
BELOW: *Patent 'Balace' one-way chilled plough by Ransomes, Sims and Jeffries of Ipswich. Photographed in 1891. The manufacturers recommended this for hillside work and claimed it could turn a furrow from 12 to 15 inches wide (305 to 380 mm), 3 to 12 inches deep (75 to 305 mm), thereby completely pulverising the soil, burying all weeds or manure and leaving the land ready for immediate sowing. The steerage lever, acting on the land wheel, is moved from end to end when the plough is reversed and acts as a second handle.*

Turnover one-way plough from the 1920 catalogue of John Huxtable of Barnstaple, Devon. When a catch between the fork of the handles is released the plough bodies are free to rotate.

assembled from stocks of prefabricated components in both iron and wood, for at this time timber was still the material for beams and handles. The range included alternative styles of frame, mouldboard and beam and a further Ransome patent of 1808 arranged for the standardisation of fittings, so that different ploughs could be built up by varying the permutation of parts.

One immediate effect of this third patent was that replacement parts need no longer be individually and laboriously made but could be quickly supplied from stock and fitted to ploughs on the farm. More importantly, it marked the mounting ascendancy of the manufacturer over the local craftsman because, at no significant extra cost, Ransome was able to assemble different ploughs that conformed to the various regional types then in use. By so offering an apparently tailor-made implement that was backed up by the supply and servicing advantages of volume production, Ransome opened up the prospect of a national market for ploughs for the first time.

The changes effected by Robert Ransome were mirrored in the many iron foundries and implement works that sprang up as the nineteenth century progressed. At the same time, the capacity to assess, test and demonstrate design modifications to ploughs was improving. An instrument, consisting essentially of a coiled spring and called a dynamometer, had been developed by the Royal Society of Arts in the 1780s for measuring the draught power requirement of ploughs. This became the standard empirical means of comparing the efficiency of ploughs one against the other. Local ploughing matches, dating back also to at least the 1780s, publicly demonstrated the effectiveness of new ploughs as well as the prowess of the ploughman.

The formation of the Royal Agricultural Society in 1839 led to the annual, peripatetic show, where competing manufacturers displayed their latest products before a discerning audience and committed them to the scrutiny of the Society's judges in organised trials. By exhibiting also at the smaller, regional shows and by appointing agents to whom equipment could be swiftly distributed through the developing railway network, it was possible for the successful plough manufacturer to become a household name in farming circles all over Britain.

The industrialisation of plough manufacture did not lead to any reduction in the number of types available because variations in soil texture and operational method still demanded differing designs. There was, however, a measure of standardisation as the important characteristics of the major types became more widely accepted and understood. Early in the nineteenth century, for example, the principle emerged of matching ploughs to seasonal requirements. So on the one hand there was the *lea, long plate* or *common plough* with a long, slightly convex mouldboard of gentle curvature designed to turn a continuous unbroken furrow in the autumn months for the weathering of soil over winter. By contrast, the *digger plough* was intended for spring work and had a short, abrupt and concave mouldboard to cause disintegration of the furrow slice and leave the soil in a pulverised state ready for final seed bed preparation.

Local conditions and personal preference might dictate whether the plough concerned was used with or without wheels. In ordinary circumstances, the wheeled variety became the more general choice and would produce sound work provided that it was correctly set and adjusted at the beginning. The wheel-less swing plough was cheaper to buy and would commend itself on soils where wheels were likely to clog or where the presence of obstacles beneath the surface called for frequent alterations of depth. Although the ploughman had through the handles more immediate control over the swing plough, a higher level of skill and stamina on his part were necessary as a result.

The ability to turn all the furrows in the same direction, so-called one-way ploughing, could have a number of benefits. It was a useful method for working narrow strips of land, while in more conventionally shaped fields there was less trampling of headlands and no need to set ridges. In the second half of the century in particular, manufacturers experimented extensively with the principle and various practical alternatives were devised. One version, the *balance plough,* was easy to operate and reduced headlands to the minimum but found its most widespread application with steam rather than horse power. Most popular was the *turnover plough,* which achieved practically its final form with the patent by John Huxtable in 1889. Two sets of share and mouldboard, above and below the beam, were alternated by the whole plough being tilted on to, and revolving about, a stock projecting from the land side.

From the closing decades of the nineteenth century, new developments were rather more concerned with the materials of manufacture than with the basic elements of design. Timber was much less commonly used although the traditional oak or ash beam was still sometimes preferred because it was both cheaper and lighter than wrought iron. Technological advances made steel, with its combination of lightness and durability, more freely available and to some extent it replaced cast iron as the favoured material for mouldboards. Horse plough evolution had effectively reached its peak by the beginning of the twentieth century. While the horse plough era still had some way to go yet, henceforth most of the innovative energies were absorbed by the new challenge of tractor ploughing.

Pair of oxen with an Oliver 'Patent Chilled Plow' c1880. These ploughs were developed by James Oliver of Indiana, USA, in the mid 1850s. It was a simple, light design suitable for most soil conditions and so well balanced that little use of the handles was necessary. Many Oliver ploughs were imported and distributed by British agents in the late nineteenth and early twentieth centuries.

Norfolk plough, c1880. This plough was well adapted to the light lands of the county and, as here, could be managed with a single handle. Characteristic of this type of plough was the steeply pitched beam resting on a gallows having wheels of equal diameter that were set close to the point of the share.

PLOUGH OPERATION

Good ploughing depends to a large degree upon correct adjustment of the different parts so that all are acting in harmony. Any discrepancies will give untidy results and increase the level of draught power required. By pitching the share downwards slightly, so that its point projects beneath the level of the sole, the plough is given a firm grip of its work. The precise degree of pitch varies according to the soil type and is found by experiment in the field.

The point of the coulter is normally sited a little above the share point but its angle of inclination depends on the work. Stiffer soils are more difficult to penetrate and call for a more forward slope from the coulter. This is the case also where there is heavy weed growth, for otherwise the root systems will snag round the coulter and impede the plough. *Disc coulters* appear to have been introduced from Holland but have been in Britain since at least the middle of the seventeenth century, when they were described by the agricultural writer Walter Blith. They can be used either in addition to or instead of the standard knife coulter and are especially good for cutting through tough, stringy surface matter. A third type, the *skim coulter,* can be positioned ahead of the knife coulter. Its function is to pare off a corner of the slices to make them bed down flatter and reduce the likelihood of weeds sprouting up between them.

On those ploughs that have them, proper attention to the setting of the wheels is most important. As well as the depth of working being affected by the height of the land wheel, the distance between the rim of the larger furrow wheel and the edge of the coulter dictates the width of furrow. The arrangement of the horse team is also a factor governing working depth. The closer the team are

Nineteenth-century Huntingdon plough with disc coulter, wooden mouldboard and Ransomes share (MERL collection). The draught rod seen at the right, and running on through a specially designed hake, was intended to relieve stress on the beam by siting the point of draught further back along its length. James Small outlined the theory of this practice in his work 'The Plough' of 1784. Later it was argued that sideways pressure, which the rod or chain did nothing to relieve, was greater than stress caused by the draught. Nevertheless, some ploughs continued to incorporate this feature.

hitched to the plough, the more steeply angled is the line of draught and the deeper the ploughing. By lengthening the traces or plough chains the angle is lowered and shallower work results. Vertical adjustment on the hake at the end of the beam allows the angle of draught, and therefore the depth, to be varied without changing the length of the chains.

The *combined hake,* in use since the eighteenth century, gives lateral adjustment as well so that the plough can be made to run straight ahead in spite of the pull from the team being slightly off-centre.

Although between 5 and 7 inches (130 to 180 mm) has generally been considered to be the standard depth of horse ploughing, other methods were adapted for particular purposes. A skimming depth of 3 or 4 inches (about 75 to 100 mm) was suitable for turning in stubble, while a deeper working of 8 to 10 inches (about 200 to 250 mm) could penetrate the pan that forms beneath the normal ploughing level to facilitate the root action of plants and allow easier access for moisture.

For conventional laying up of the land in seams using the common plough, it was established in the nineteenth century that the optimum dimensions for the width and depth of the individual slices should be in the ratio of 10 to 7. A width

At work in Essex with a wooden beamed plough and three horses in line. Wooden spreaders, positioned behind the front and middle horses, keep the trace chains away from their sides. The rear horse has a separate chain coupling to the swingletree to allow for the steeper angle of draught on the plough at this point.

ABOVE: *Plough drawn by three horses yoked abreast and followed by a furrow press. This was used especially in drier areas to break down the cavity beneath the furrow and provide a firmer bed for corn sown broadcast over the surface.*

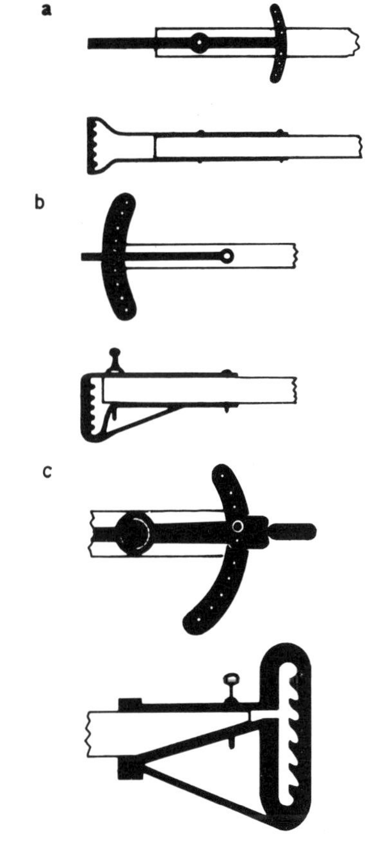

RIGHT: *The hake at the end of the beam provides the means of attachment of the draught chains to the plough. By the eighteenth century, hakes commonly in use allowed both vertical adjustment of the line of draught and also horizontal adjustment to take account, for example, of whether the horses were hitched in line or abreast. The first two drawings are taken from John Arbuthnot's plough writings of the 1770s. He preferred the Suffolk hake (b) to the common hake (a), which he said required too much effort to adjust so that the ploughman always tended to let the plough run too deep. Drawing (c) shows plan and elevation of a hake for a Ransomes plough of the late nineteenth century.*

of 10 inches (254 mm) should, therefore, mean a depth of 7 inches (178 mm). In the process of ploughing, each slice is rotated through 135 degrees so that it rests at an angle of 45 degrees against its neighbour and exposes to the air previously buried surfaces of equal length. Other shapes of slice, notably the crested variety, have periodically been advocated but the rectangular shape remained dominant for the horse plough.

A team pulling a plough cutting a 10 inch (254 mm) wide furrow will have to travel a fraction under 10 miles (16 km) in order to plough an acre (0.4 ha). At a steady pace of 1½ to 2 miles per hour (2.4 to 3.2 km/h) this will take about 6 hours continuous work or 8 hours with the normal stoppages for rest and adjustments. The amount of power required to pull a plough varies markedly with the type of soil and with such factors as whether it is wet or dry. On heavy land, the 10 inch by 7 inch (254 mm by 178 mm) furrow will need 840 pounds (381 kg) of draught, while on light land this drops to 280 pounds (127 kg). These figures represent the theoretical pulling power at 1½ miles per hour (2.4 km/h) of

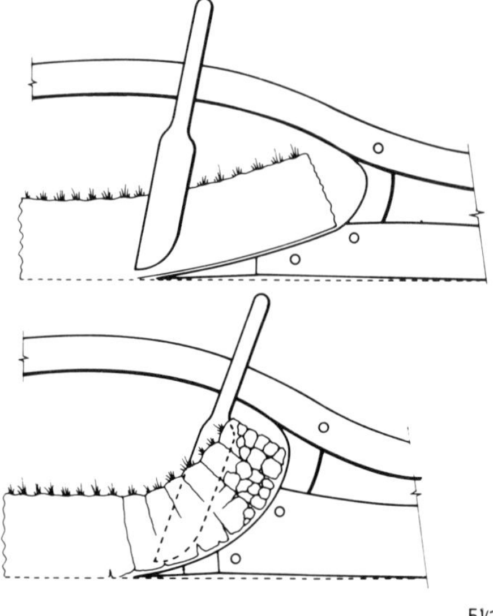

LEFT: *Top: on the lea or common plough the furrow slice is raised smoothly by share and breast so that it remains intact and the land is laid up in unbroken seams. Bottom: the digger plough has a considerably more abrupt curve of share and breast to raise the furrow slice sharply and cause it to break up. The cutting action of share and coulter occurs much closer to the breast than on the lea plough.*

BELOW: *Standard arrangement for horse ploughing with a fixed mouldboard plough. The ridges are marked out across the field with the distance between them, about 22 yards (20 m), known as a 'land'. A quarter land on either side of adjacent ridges (1 and 2 on the diagram) is then 'gathered' by turning clockwise at the headlands. The intervening space is 'cast' by working in from the outside and turning anti-clockwise at the headlands to finish at an open furrow mid-way between the two ridges.*

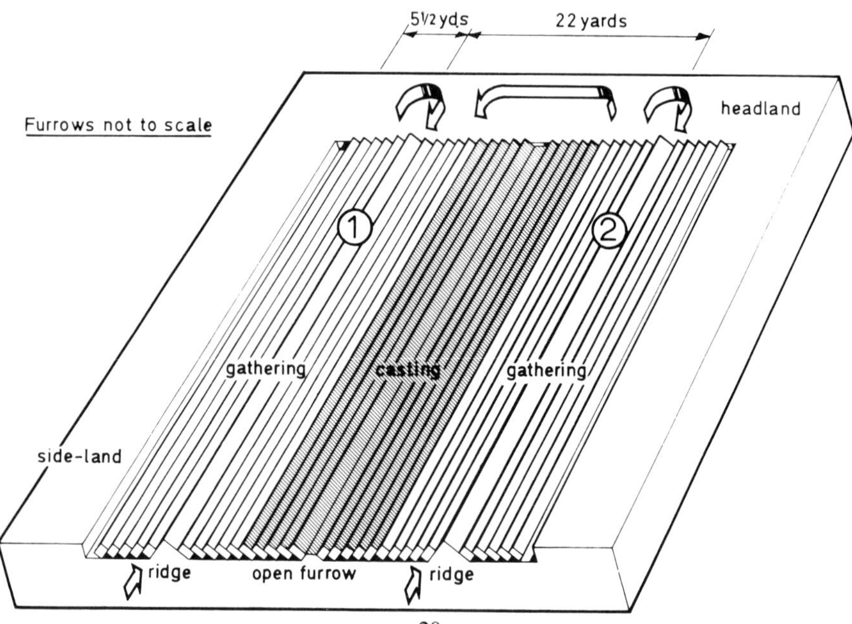

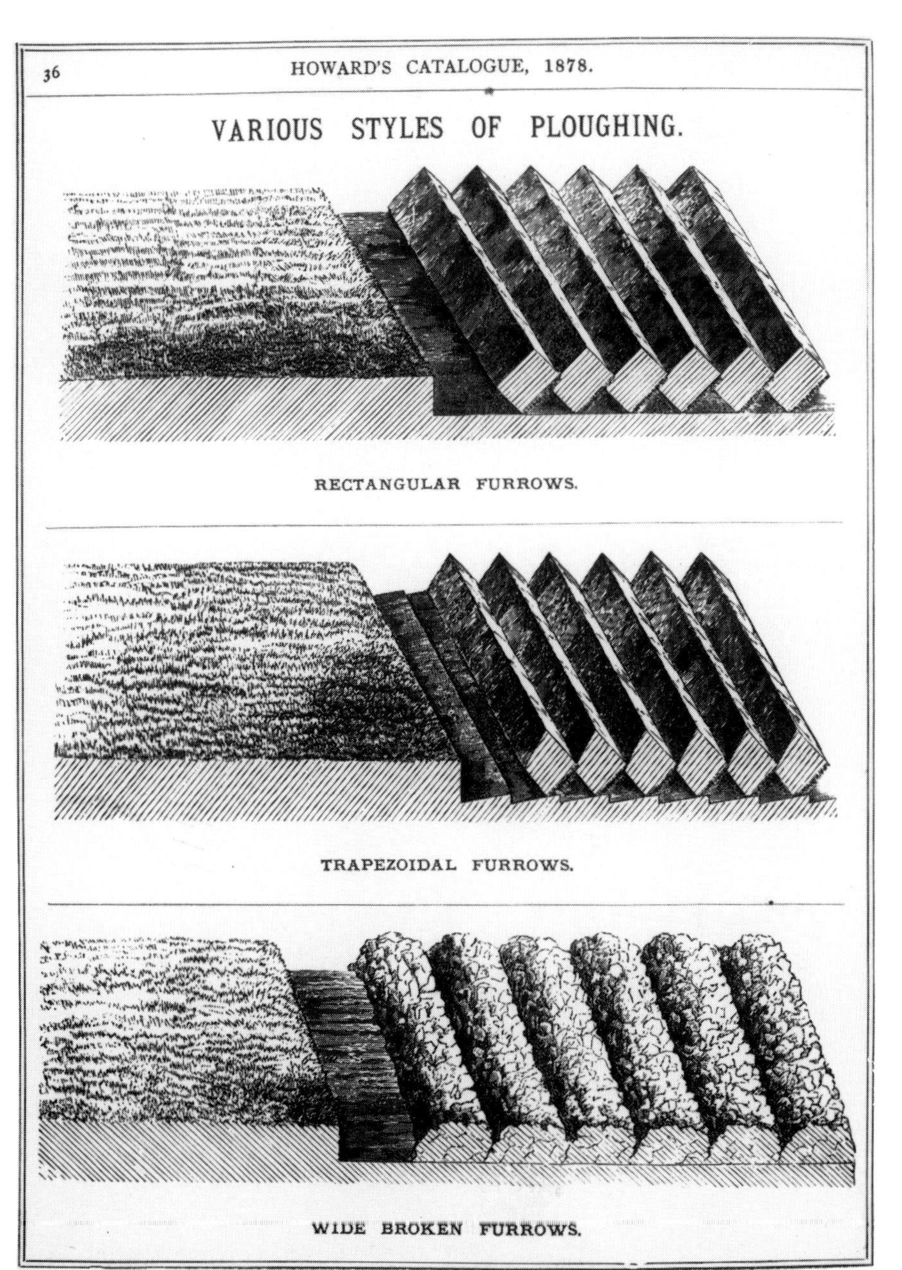

There was a long debate on the optimum shape of furrows although the conventional rectangular kind, lying at 45 degrees to the horizontal, had most support. The crested furrow — here referred to as trapezoidal — was cut by a share having a raised wing. While it exposed a greater surface to the air and provided a more ample covering for seed, it also left a larger cavity beneath the surface and an uneven furrow floor.

Turning at the headland with three horses yoked in line. This was often the method adopted for heavy wet land as the whole team then trod the firmer furrow bottom and did not trample the unploughed area.

2.5 horses in the first case and 0.8 horses in the second. Working at a faster pace leads to a decline in the pulling power per horse so that a bigger team may become necessary.

As a consequence of the common plough having a fixed mouldboard, usually on the right, it follows that a systematic method is needed if all the field is to be laid up in seams resting against each other. One answer is to plough in a continuous circle by either beginning at the edge of the field and going round and round into the middle or by starting at the centre and working outwards. General practice on the individual strips of the medieval open fields was to begin from a ridge in the middle and then plough round it, laying the slices inwards on either side. Repetition over a long period of time led to the characteristic undulating ridge and furrow effect which not only assisted with field drainage but also clearly delineated the boundaries of each strip.

Through to the twentieth century, a not dissimilar technique has been commonly used to plough an ordinary field with horses. First, the ridges or feerings are marked out across the field, dividing it into lands of about 22 yards (20 m) in width. Then the sections are worked up and down by gathering and casting to finish with an open furrow in the middle of each land. When all the lands have been completed, the headlands and sidelands can be dealt with by roundabout ploughing to the outer edge.

LEFT: *Gathering round the central ridge with a single-furrow plough, 1938. At the headland the team turns in a clockwise direction to plough back down the other side of the ridge.*

RIGHT: *These two teams are in the casting stage, turning anticlockwise at the headlands, to finish with an open furrow in the centre of the remaining area to be ploughed.*

ABOVE: *Steam ploughing apparatus, from the 1861 catalogue of Howard of Bedford. By driving each drum in turn, the steel cables were alternately wound in and paid out, and the double-furrow reversible plough pulled backwards and forwards across the field. Using an 8 or 10 horsepower portable engine, Howards claimed that 7 to 12 acres (2.8 to 4.8 ha) a day could be broken up. The main advantage of this system was its comparative cheapness, at £200 excluding the cost of the engine.*

BELOW: *Seven-furrow Fowler balance plough working on the double engine system of steam cultivation, 1912. At the end of each run, the implement was reversed to bring the opposite set of plough bodies into play.*

Motor cable tackle operating a nine-furrow balance plough in 1922. A few of these sets were manufactured by John Fowler and Company of Leeds between 1919 and 1930, mostly for export.

MECHANICAL PLOUGHING

Considerable efforts were applied from the middle of the nineteenth century to make steam-powered ploughing a practical proposition. To a limited degree these were successful and steam plough sets could still be seen at work in the 1930s but the earlier high hopes for steam cultivation in Britain were never entirely fulfilled. Direct haulage of the plough across the field was in most cases out of the question for both practical and economic reasons. Instead, the plough was hauled by means of a cable driven from an engine positioned at the side or in the corner of the field. A number of ingenious cable systems were devised but the most widely used was that employing two engines, one on either side of the field, winding a multi-furrow balance plough between them.

The principal benefit of steam power was that heavier and deeper work could be accomplished more quickly and in the correct season. This alleviated the backlog that might occur at busy times of the year when horses alone were used and meant that more of the land was ready to be sown with corn in the autumn months, rather than the following spring, with a possible improvement in yields. The necessary capital outlay on equipment, however, was very great, especially as engines of such power were not required in other farm operations. For this reason, few farmers purchased their own ploughing sets but rather relied on the services of steam contractors.

With the appearance of the internal combustion engine, motor-cable haulage equipment was also built but, more importantly, this new, lighter form of power made direct traction feasible. Early tractor ploughing, of the First World War period, required two-man operation with, in addition to the driver, another man sitting on the plough itself to steer it and work the depth levers. Although these steerage ploughs were still being manufactured in the 1920s, a self-lift alternative controlled by the tractor driver alone was available in 1919 and became common in the inter-war years.

One-man operation made the tractor-drawn plough a tough competitor for the

ABOVE: *A Fowler six-furrow gang plough drawn by a Fowler 50 brake horsepower internal combustion engined road locomotive, September 1912.*

BELOW: *International Titan 10-20 tractor with a Ransomes RYLT four-furrow riding plough. These ploughs were first introduced in 1914 and by the end of the First World War almost two thousand Titans had been imported from America to help boost domestic food production.*

Land Army girl with Fordson tractor and two-furrow Oliver plough near Barnet, 1944. No second operator was needed on these trailed ploughs because the levers or screws controlling the depth could be reached from the driver's seat. Clearly visible by the land wheel is the ratchet mechanism which cranked the axle arm up to a more vertical position and raised the plough bodies out of the soil. On reaching the end of the furrow the driver pulled the trip lever (positioned above the depth screw) to operate the mechanism and then pulled it a second time to release the ratchet when the plough was in position for the return run.

horse team. When travelling at around 2 miles per hour (3.2 km/h), for example, a three-furrow plough could cover between 3 and 5 acres (1.2 to 2.0 ha) in a day. This enabled work to advance rapidly after harvest when the soil was dry and likely to benefit more. Against this, early tractors were not altogether reliable so considerable time might be lost through mechanical failure. Wheel slippage caused problems on softer soils before pneumatic tyres were developed in the 1930s.

In the same decade the tractor and plough were merged into a combined controllable unit following the innovative work of Harry Ferguson. His system employed a three-point linkage at the rear of the tractor on to which a wheelless plough or other implement could be hitched and through which the depth, draught and raising could be manipulated hydraulically by the driver. Tractors of this kind were first offered commercially by David Brown in 1936 and the underlying principles have remained standard ever since. Post-war developments have refined draught control into a response control for variable working conditions and added a position control for precise adjustment of the implement relative to the tractor.

Modern tractor plough bodies are normally fixed to a diagonal main beam, the angle of which can be altered to change the width of furrow. The larger ploughs, up to twelve-furrow size, are semi-mounted but hydraulic action can still control the draught and, by working through a rear wheel, lift the bodies in

ABOVE: *Stubble ploughing with a Ferguson tractor and three-furrow plough, Sotherton, Somerset, 1953. With the mounted plough, tractor and implement operated as a single unit. Tractors incorporating Harry Ferguson's hydraulic and three-point linkage systems were built by David Brown Tractors Ltd from 1936 to 1939 and Ford of America from 1939 to 1947. Mass production of the Ferguson TE20 began in Coventry in 1946 under the Standard Motor Company. A merger in 1953 produced a new company: Massey-Harris-Ferguson Ltd.*

LEFT: *The TS70 four-furrow mounted plough developed by Ransomes in 1953 for use on the Fordson Major tractor. This plough was strong enough for medium land yet could be lifted by a tractor and easily transported. A steel wheel, adjustable from the driver's seat, maintained an even depth while the width was fixed at 10 inches (250 mm).*

and out of work.

Traditional methods of ploughing, using ridges and open furrows to lay the field up in lands, continued from the horse era into the tractor age and may still be seen. The principal distinction early on was that, whereas a horse team could plough up and down on either side of the same ridge, a tractor, with a larger turning circle, could not. A revised pattern was therefore adopted with the tractor working on different pairs of ridges in sequence. Other systems of ploughing have since been evolved and are now widely employed, many of them based on the square ploughing principle. A common version requires a small land to be ploughed in the middle of the field so that the remaining area is of equal breadth on all four sides. This is then ploughed out round and round to the edge. Modern technology, however, has made the reversible plough, with two sets of bodies mounted one above the other, a popular choice for reasons of speed and simplicity of working.

ABOVE: *A mounted three-furrow reversible plough by David Brown Tractors Ltd, 1963. At each headland, the pull of a lever was sufficient to turn the plough over and bring the opposite set of bodies into work. A reversible plough is more expensive but it makes for quicker operation because there are no ridges or open furrows and time spent turning at the headlands is much reduced.*
BELOW: *A Roadless Ploughmaster tractor with a Doe eight-furrow semi-mounted plough, 1964. Larger ploughs, normally from five furrows upwards, are usually arranged in this way with hydraulic control operating at both front and rear.*

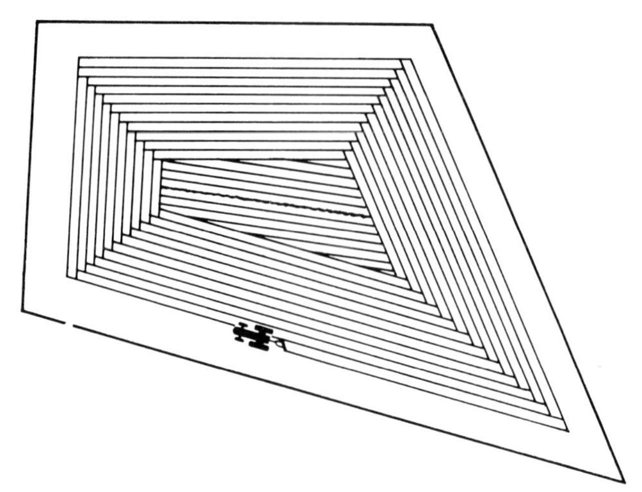

ABOVE: *A common method of ploughing by tractor that avoids setting the field up in ridges. A small land is ploughed in the centre of the field so that the distance to the outer edge is equal on all four sides. The tractor then ploughs the remainder round and round to the periphery.*

BELOW: *An 'Iron Horse' tractor with single-furrow plough manufactured by the British Anzani Engineering Company Ltd, of Middlesex. It has a 6 horsepower engine and is fitted with Rotaped tracklayer wheels by Leeford of London. Small tractors like this, with a full complement of accompanying implements, were made by a number of firms and were popular during the 1940s and 1950s for market garden work. Depth of ploughing is regulated by the crank handle visible above the mouldboard.*

PLACES TO VISIT

Intending visitors are advised to find out the times of opening before making a special journey.

Acton Scott Historic Working Farm, Wenlock Lodge, Acton Scott, Church Stretton, Shropshire SY6 6QN. Telephone: 01694 781306.

Ashwell Village Museum, Swan Street, Ashwell, Baldock, Hertfordshire SG7 5NY. Telephone: 01462 742956.

Beamish, The North of England Open Air Museum, Beamish, County Durham DH9 0RG. Telephone: 01207 231811.

Beck Isle Museum of Rural Life, Beck Isle, Pickering, North Yorkshire YO18 8DU. Telephone: 01751 473653.

Bodmin Farm Park, Fletchers Bridge, Bodmin, Cornwall. Telephone: 01208 827074.

Breamore Countryside Museum, Breamore House, Breamore, near Fordingbridge, Hampshire SP6 2DF. Telephone: 01725 512468.

Cambridge and County Folk Museum, 2/3 Castle Street, Cambridge CB3 0AQ. Telephone: 01223 355159.

Castle Cary Museum, The Market House, Castle Cary, Somerset. Telephone: 01963 350462.

Clitheroe Castle Museum, Castle Hill, Clitheroe, Lancashire BB7 1BA. Telephone: 01200 424635.

Cogges Manor Farm Museum, Church Lane, Cogges, Witney, Oxfordshire OX8 6LA. Telephone: 01993 772602.

Cotswold Countryside Collection, Northleach, Gloucestershire GL54 3JH. Telephone: 01451 860715.

Craven Museum, Town Hall, High Street, Skipton, North Yorkshire BD23 1AH. Telephone: 01756 794079.

Dales Countryside Museum, Station Yard, Hawes, North Yorkshire DL8 3NT. Telephone: 01969 667494.

Dorset County Museum, High West Street, Dorchester, Dorset DT1 1XA. Telephone: 01305 262735.

Elvaston Castle Estate Museum, Elvaston Castle Country Park, Borrowash Road, Elvaston, Derbyshire DE72 3EP. Telephone: 01332 573799.

Felin Puleston Agricultural Museum, Erddig Estate, Wrexham, Clwyd LL13 0YT. Telephone: 01978 263241.

Finch Foundry, Sticklepath, Okehampton, Devon EX20 2NW. Telephone: 01837 840046.

Folk Museum of West Yorkshire, Shibden Hall, Halifax, West Yorkshire HX3 6XG. Telephone: 01422 352246.

Guernsey Farm Museum, Castel, Guernsey GY5 7UJ. Telephone: 01481 55384.

Gwent Rural Life Museum, The Malt Barn, New Market Street, Usk, Monmouth NP5 1AU. Telephone: 01291 673777.

Highland Folk Museum, Duke Street, Kingussie, Inverness-shire PH21 1JG. Telephone: 01540 661307.

James Countryside Museum, Bicton Park, East Budleigh, Budleigh Salterton, Devon EX9 7DP. Telephone: 01395 568465.

Lackham Agricultural Museum, Lackham College, Lacock, Chippenham, Wiltshire SN15 2NY. Telephone: 01249 443111.

Long Shop Museum, Main Street, Leiston, Suffolk IP13 4ES. Telephone: 01728 832189.

Melton Carnegie Museum, Thorpe End, Melton Mowbray, Leicestershire LE13 1RB. Telephone: 01664 69946.

Milton Keynes Museum of Industry and Rural Life, Southern Way, Wolverton, Milton Keynes, Buckinghamshire MK12 5EJ. Telephone: 01908 316222.

Museum of East Anglian Life, Stowmarket, Suffolk IP14 1DL. Telephone: 01449 612229.

Museum of Kent Life, Lock Lane, Sandling, Maidstone, Kent ME14 3AU. Telephone: 01622 763936.

Museum of Lakeland Life and Industry, Abbot Hall, Kendal, Cumbria LA9 5AL. Telephone: 01539 722464.

Museum of Lincolnshire Life, The Old Barracks, Burton Road, Lincoln LN1 3LY. Telephone: 01522 528448.

Museum of Welsh Life, St Fagans, Cardiff CF5 6XB. Telephone: 01222 569441.

Norfolk Rural Life Museum and Union Farm, Beech House, Gressenhall, East Dereham, Norfolk NR20 4DR. Telephone: 01362 860563.

Normanby Park Farming Museum, Normanby Hall Country Park, Scunthorpe DN15 9HU. Telephone: 01724 720588.

Norris Museum, The Broadway, St Ives, Huntingdon, Cambridgeshire PE17 4BX. Telephone: 01480 465101.

North Cornwall Museum and Gallery, The Clease, Camelford, Cornwall PL32 9PL. Telephone: 01840 212954.

Oxfordshire County Museum, Fletcher's House, Woodstock, Oxfordshire OX20 1SN. Telephone: 01993 811456.

Rural History Centre (incorporating Museum of English Rural Life), The University, Whiteknights, Reading, Berkshire RG6 6AG. Telephone: 0118-931 8660.

Rural Life Centre, Old Kiln Museum, Reeds Road, Tilford, Farnham, Surrey GU10 2DL. Telephone: 01252 795571.

Rutland County Museum, Catmos Street, Oakham, Rutland LE15 6HW. Telephone: 01572 723654.

Ryedale Folk Museum, Hutton-le-Hole, North Yorkshire YO6 6UA. Telephone: 01751 417367.

Science Museum, Exhibition Road, South Kensington, London SW7 2DD. Telephone: 0171-938 8000.

Scolton Manor Museum, Spittal, Haverfordwest, Pembrokeshire SA62 5QL. Telephone: 01437 731328.

Scottish Agricultural Museum, Ingliston, by Edinburgh EH28 8NB. Telephone: 0131-333 2674.

Shugborough Estate, Milford, near Stafford ST17 0XB. Telephone: 01889 881388.

Somerset Rural Life Museum, Abbey Farm, Chilkwell Street, Glastonbury, Somerset BA6 8DB. Telephone: 01458 831197.

Swaledale Folk Museum, Reeth Green, Reeth, near Richmond, North Yorkshire DL11 6QT. Telephone: 01748 884373.

Tiverton Museum, St Andrew Street, Tiverton, Devon EX16 6PH. Telephone: 01884 256295.

Upminster Tithe Barn Agricultural and Folk Museum, Hall Lane, Upminster, Essex RM14 1AU. Telephone: 01708 447535.

Weald and Downland Open Air Museum, Singleton, Chichester, West Sussex PO18 0EU. Telephone: 01243 811348.

Yorkshire Museum of Farming, Murton Park, York YO1 3UF. Telephone: 01904 489966.

FURTHER READING

Blith, Walter. *English Improver Improved.* 1653.

Board of Agriculture. *General View of the Agriculture* (of each county). 1793-1819.

Bond, J. R. *Farm Implements and Machinery.* 1923.

Bonnett, Harold. *Saga of the Steam Plough.* 1965.

Brown, J. *Farm Machinery 1750-1945.* 1989.

Culpin, C. *Farm Machinery.* Tenth edition, 1981.

Fussell, G. E. *The Farmer's Tools.* 1952; new edition 1981.

Passmore, J. B. *The English Plough.* 1930.

Ransome, J. Allen. *The Implements of Agriculture.* 1843.

Rees, Sian. *Agricultural Implements in Prehistoric and Roman Britain* (part i). 1979.

Roworth, E. J. *Good Ploughing.* 1977.

Royal Agricultural Society of England. *Annual Journals.*

Slight, James, and Burn, R. Scott. *The Book of Farm Implements and Machines.* 1868.

Small, James. *Treatise on Ploughs and Wheel Carriages.* 1784.

Spence, Clark. *God Speed the Plow.* 1960.